3 Arizona AASA Grade 3 Math Practice Tests

Full-Length Test Prep with Detailed Answer Explanations

Dr. A. Nazari

3 Practice Tests to Get You Started!

Hey there, future math whiz! ⭐

This book has **3 full practice tests** to help you warm up for the real thing. Think of it like stretching before a big game — these tests will get your brain ready and show you what to expect!

 Three tests is the **perfect start**!

 Each one helps you feel **more ready**!

👍 You'll be surprised how much you **already know**!

Sharpen your pencil and let's get warmed up! 👌

> ❝ Three practice tests is a great way to start. Take your time with each one, and you'll feel more confident every step of the way! ❞

📘 How to Use This Book 📘

Your quick-start guide to 3 great practice tests!

📋 What's Inside This Book

- **3 Full-Length Practice Tests** — Each one covers all the Grade 3 math topics you need to know!

- **Answer Key with Explanations** — Find out why each answer is correct, not just what the answer is.

- **Reference Pages** — A math symbols chart and multiplication table you can peek at any time.

- **A Test Tracker** — Write down your scores and watch your confidence grow!

📅 A Simple 3-Test Plan

With just 3 tests, here's a great way to use them:

- **Test 1** — **The Warm-Up.** Take this test without a timer. Get comfortable with the question types. Don't worry about your score — just do your best!

- **Test 2** — **The Practice Round.** After reviewing Test 1, try this one with a timer (ask a grown-up!). Focus on the topics that were tricky last time.

- **Test 3** — **The Real Deal.** Treat this like the actual test: quiet room, timed, no peeking at answers. See how much you've improved!

⚪ Multiple Choice

Pick the **one best answer** from choices A, B, C, or D. Not sure? Cross out the ones you know are wrong, then pick from what's left. That's a smart move!

✏️ Short Answer

Write your answer **and** show your work! Even if your final answer isn't right, showing your steps can earn you credit. Use scratch paper if you need more room.

66 After Each Test 99

Flip to the Answer Key and check your work. For every question you got wrong, **read the explanation carefully.** Then write the tricky topics on your Test Tracker page. If you need extra help, grab our **Grade 3 Math Study Guide!**

⭐ **Fun fact:** *Three tests is all it takes to see real improvement! Most kids feel way more confident after just a few rounds of practice.* ⭐

Find more at
ViewMath.com/AZ-Grade3

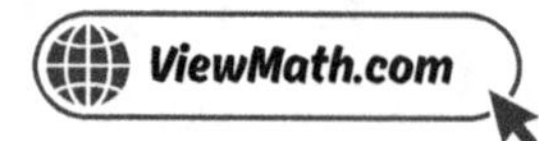

 Tips for Test Day

Easy tricks to help you feel calm and do your best!

☾ The Night Before

- ✅ **Sleep early** — *your brain learns while you sleep!*
- ✅ **Pack your supplies** — *pencils, eraser, scratch paper, all ready to go.*
- ✅ **Tell yourself:** *"I've been practicing. I'm going to do great!"*

👍 5 Simple Rules for Every Test

1. **Read the question twice.** *The first time to understand it. The second time to catch details.*
2. **Show your work.** *Write the steps down, even on scratch paper. It helps you think!*
3. **Skip the hard ones.** *Put a small star next to tricky questions and come back later. Answer the easy ones first!*
4. **Never leave a blank.** *For multiple choice, your best guess is better than no answer at all.*
5. **Check your work.** *Finished early? Go back and re-read your answers.*

✅ Smart Moves

- Take a deep breath before you begin
- Underline key words in the question
- Use drawings or number lines to help
- Cross out wrong answers first
- Double-check addition and subtraction

❌ Traps to Avoid

- Rushing and not reading carefully
- Picking the first answer that "looks right"
- Forgetting to carry or borrow numbers
- Skipping a question permanently
- Panicking when you see a tough problem

> ❝ *Remember, the very first practice test is the hardest — not because the questions are harder, but because everything is new! By Test 3, you'll feel like a pro. Trust me!* ❞

🧰 Get Ready to Practice 🧰

Here's everything you need before you start!

Pencils

Sharpened and ready!

Eraser

Everyone makes mistakes!

Scratch Paper

For working things out

A Calm Spot

Somewhere quiet to focus

A Grown-Up

To help set a timer

A Can-Do Attitude

You've totally got this!

✔ Allowed During Tests

- Pencils and erasers
- Blank scratch paper
- The **reference pages** in this book
- A ruler (for measurement questions)

✖ Not Allowed

- Calculators
- Phones, tablets, or computers
- Help from anyone else
- Your study guide (save it for after!)

👥 For Parents & Teachers

- With only 3 tests, **space them at least a week apart.** This gives time to review mistakes before trying the next one.
- Let your child take Test 1 untimed to build familiarity.
- After each test, go through the Answer Key together. Focus on **understanding the "why,"** not just the score.
- If a topic keeps tripping them up, review it in our **Grade 3 Math Study Guide** before the next practice test.
- Celebrate every bit of progress — even getting one more question right is a win!

X^1 Math Reference Sheet X^1

Symbol	Name	What It Means	
$+$	Plus (Add)	Put numbers together.	$3 + 5 = 8$
$-$	Minus (Subtract)	Take away from a number.	$9 - 4 = 5$
$\times$	Times (Multiply)	Add equal groups.	$4 \times 3 = 12$
$\div$	Divide	Split into equal groups.	$12 \div 3 = 4$
$=$	Equals	Both sides are the same.	$2 + 3 = 5$
$>$	Greater Than	The left number is bigger.	$7 > 3$
$<$	Less Than	The left number is smaller.	$2 < 9$
$\frac{1}{2}$	Fraction Bar	Part of a whole.	$\frac{1}{2}$ means 1 out of 2 equal parts

📖 Key Math Words

- **Sum** — the answer when you add
- **Difference** — the answer when you subtract
- **Product** — the answer when you multiply
- **Quotient** — the answer when you divide
- **Factor** — a number you multiply
- **Array** — objects in rows and columns
- **Fraction** — a part of a whole

- **Numerator** — the top number in a fraction
- **Denominator** — the bottom number
- **Equation** — a math sentence with $=$
- **Estimate** — a smart guess, close to the real answer
- **Perimeter** — the distance around a shape
- **Area** — the space inside a shape
- **Rounding** — making a number simpler by going to the nearest ten or hundred

🔍 *Word Problem Clue Words*

- **Add** (+): *in all, total, altogether, combined, sum, both, more*

- **Subtract** (−): *how many more, how many left, fewer, difference, remain*

- **Multiply** (×): *each, every, groups of, times, rows of, per*

- **Divide** (÷): *share equally, split, each group, how many groups, per*

Find more at
ViewMath.com/AZ-Grade3

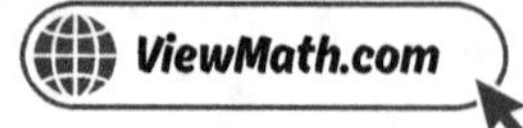

▦ Multiplication Table ▦

×	1	2	3	4	5	6	7	8	9	10	11
1	1	2	3	4	5	6	7	8	9	10	11
2	2	4	6	8	10	12	14	16	18	20	22
3	3	6	9	12	15	18	21	24	27	30	33
4	4	8	12	16	20	24	28	32	36	40	44
5	5	10	15	20	25	30	35	40	45	50	55
6	6	12	18	24	30	36	42	48	54	60	66
7	7	14	21	28	35	42	49	56	63	70	77
8	8	16	24	32	40	48	56	64	72	80	88
9	9	18	27	36	45	54	63	72	81	90	99
10	10	20	30	40	50	60	70	80	90	100	110
11	11	22	33	44	55	66	77	88	99	110	121

♀ How to Use This Table

To find **4 × 7**:

1. Find **4** in the left column (blue).
2. Find **7** in the top row (blue).
3. Follow the row and column until they meet: the answer is **28**!

📈 My Confidence Tracker 📈

Record your scores below. You'll be amazed at your progress!

My name: _______________________________

☑ Test	📅 Date	⭐ Score	🙂 How I Feel
1		/	
2		/	
3		/	

💡 My Quick Reflection

The easiest topic for me was:

____

The trickiest topic for me was:

____

One thing I got better at from Test 1 to Test 3:

____

Next time I want to try:

____

> *You just finished 3 practice tests — that's awesome! Compare your first score to your last. I bet you'll see real improvement. Ready for more? Check out our 5-test or 7-test books for even more practice!*

Find more at
ViewMath.com/AZ-Grade3

ViewMath.com

 Table of Contents

Here's what we'll explore together!

⭐ Practice Test 1 1

⭐ Practice Test 2 9

⭐ Practice Test 3 17

⭐ Answer Key & Explanations 25

 Let's learn and have fun!

Practice Test 1

 *30 Questions*

 Before You Start 

- ✓ **Read each question carefully** before choosing your answer.
- ✓ **Show your work** on scratch paper when you need to.
- ✓ **Skip hard questions** and come back to them later.
- ✓ **Check your answers** when you're done.
- ✓ **Take your time** — there's no rush!

 You've Got This! 

Do your best and show what you know!

1. Which number has a 4 in the hundreds place and a 7 in the thousands place?

   (A) 4,731

   (B) 7,431

   (C) 7,341

   (D) 3,470

2. Which number rounds to 500 when rounded to the nearest 100?

   (A) 428

   (B) 449

   (C) 462

   (D) 551

3. Which group contains only even numbers?

   (A) 12, 34, 57

   (B) 20, 46, 88

   (C) 31, 50, 72

   (D) 14, 63, 90

4. Marcus says $276 + 358 = 534$. What mistake did he most likely make?

   (A) He forgot to regroup in the ones column

   (B) He forgot to regroup in the tens column

   (C) He added the hundreds digits incorrectly

   (D) He subtracted instead of adding

5. If $6{,}412 - 2{,}785 = 3{,}627$, which addition problem checks this answer?

   (A) $6{,}412 + 2{,}785$

   (B) $3{,}627 + 6{,}412$

   (C) $3{,}627 + 2{,}785$

   (D) $2{,}785 - 3{,}627$

6. Estimate $248 + 175 + 362$ by rounding each number to the nearest hundred.

   Your Answer:

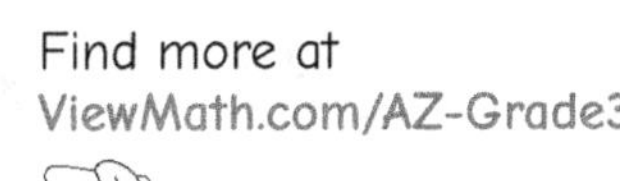
Find more at
ViewMath.com/AZ-Grade3

7. Each car has 4 tires. How many tires are on 3 cars?

Your Answer:

8. Which of these numbers is a multiple of both 6 and 8?

(A) 36

(B) 42

(C) 48

(D) 56

9. I am a number. When you divide me by 8, you get 9. What number am I?

Your Answer:

10. Find the missing number. $72 \div ? = 8$

(A) 7

(B) 8

(C) 9

(D) 10

11. Odd × odd = ?

(A) Always even

(B) Always odd

(C) Sometimes even, sometimes odd

(D) Always 0

12. What is $9 + 18 \div 3$?

(A) 9

(B) 12

(C) 15

(D) 27

Find more at
ViewMath.com/AZ-Grade3

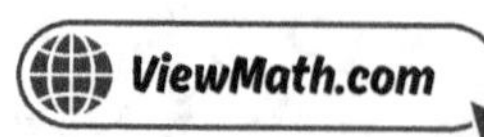

13. You eat $\frac{2}{6}$ of a cake. How many equal parts does the whole cake have?

Your Answer:

14. What whole number does $\frac{6}{3}$ equal?

(A) 1

(B) 2

(C) 3

(D) 6

15. You are at $\frac{3}{6}$ on a number line. You hop 2 more times by $\frac{1}{6}$. Where do you land?

Your Answer:

16. $\frac{3}{4} = \frac{?}{8}$. What is the missing numerator?

(A) 3

(B) 4

(C) 6

(D) 7

17. Which of these fractions equals 0?

(A) $\frac{1}{1}$

(B) $\frac{0}{4}$

(C) $\frac{4}{4}$

(D) $\frac{4}{0}$

18. Is $\frac{2}{6}$ more or less than $\frac{1}{2}$?

Your Answer:

Find more at
ViewMath.com/AZ-Grade3

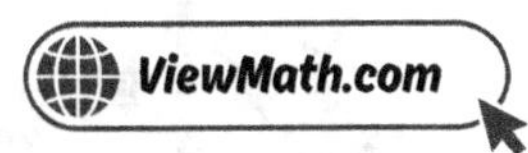

19. The minute hand points to the 12 and the hour hand points to the 5. What time is it?

(A) 12:05

(B) 12:25

(C) 5:12

(D) 5:00

20. A string is $6\frac{1}{4}$ inches long. A ribbon is $6\frac{3}{4}$ inches long. How much longer is the ribbon than the string?

(A) $\frac{1}{4}$ inch

(B) $\frac{1}{2}$ inch

(C) $\frac{3}{4}$ inch

(D) 1 inch

21. A bag of apples has a mass of 3 kg. A bag of oranges has a mass of 5 kg. What is the total mass?

(A) 2 kg

(B) 15 kg

(C) 8 kg

(D) 35 kg

22. A large milk jug holds 4 liters. You pour out 1 liter for cereal. How much milk is left?

Your Answer

23. How much is a quarter worth?

(A) 1 cent

(B) 5 cents

(C) 10 cents

(D) 25 cents

24. You buy a book for $4.50 and pay with a $10 bill. How much change do you get?

(A) $4.50

(B) $5.50

(C) $6.50

(D) $14.50

Find more at
ViewMath.com/AZ-Grade3

25. A bar graph has a scale that counts by 5s. A bar reaches up to the 4th line on the scale. What value does the bar show?

(A) 4

(B) 9

(C) 15

(D) 20

26. A line plot shows the lengths of 10 crayons. The data is shown below:

2 in.	$2\frac{1}{2}$ in.	3 in.	$3\frac{1}{2}$ in.	4 in.
2	3	4	1	0

How many crayons are shorter than 3 inches?

(A) 2

(B) 4

(C) 5

(D) 9

27. A shape has 4 sides. Two sides are 8 cm long and two sides are 3 cm long. It has 4 right angles. What shape is it?

Your Answer:

28. A square has sides that are 7 inches long. What is its area?

(A) 14 sq in

(B) 28 sq in

(C) 42 sq in

(D) 49 sq in

29. Lily says a rectangle that is 5 cm long and 3 cm wide has a perimeter of 15 cm. What mistake did she make?

(A) She subtracted instead of adding.

(B) She found the area instead of the perimeter.

(C) She forgot to add all 4 sides.

(D) She divided instead of multiplying.

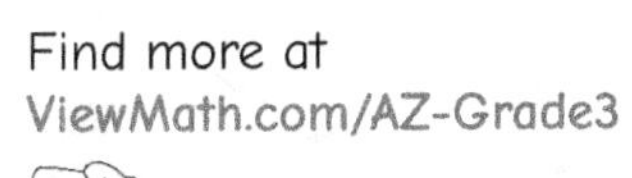
Find more at
ViewMath.com/AZ-Grade3

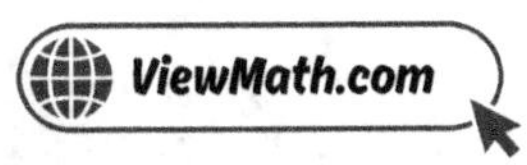

30. If one pizza is cut into 4 equal slices and another identical pizza is cut into 8 equal slices, which slices are bigger?

(A) The $\frac{1}{8}$ slices

(B) The $\frac{1}{4}$ slices

(C) They are the same size.

(D) You cannot tell.

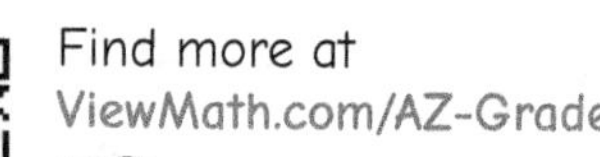

⭐ *End of Practice Test 1* ⭐

Great job finishing the test!

 My Score

I got ______________ out of 30 questions right.

*Check your answers in the **Answer Key** at the back of the book.*

💡 *Review any questions you missed. That's how we learn!*

📊 Check Your Score Online!

Visit **ViewMath Academy** to enter your answers and see which topics you need to review. You can also explore lessons, take quizzes, track your scores, and save your progress!

viewmath.com/score/3.1.AZ.01

Or go to *viewmath.com/score* and enter code: 3.1.AZ.01

2

Practice Test 2

 30 Questions

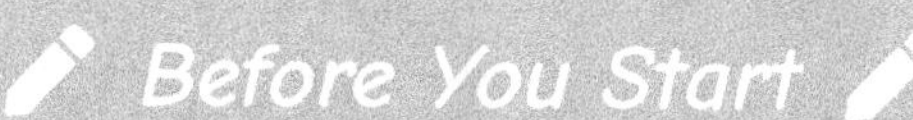

✏ Before You Start ✏

- ✓ **Read each question carefully** before choosing your answer.
- ✓ **Show your work** on scratch paper when you need to.
- ✓ **Skip hard questions** and come back to them later.
- ✓ **Check your answers** when you're done.
- ✓ **Take your time** — there's no rush!

 You've Got This! 

Do your best and show what you know!

1. A number has 3 thousands, 7 hundreds, 0 tens, and 5 ones. What is the number?

Your Answer:

2. Maria rounded 438 to the nearest 10 and got 430. Is she correct?

(A) No, it should be 440

(B) Yes, 438 rounds to 430

(C) No, it should be 400

(D) No, it should be 450

3. List all the even numbers between 21 and 30.

Your Answer:

4. What is $198 + 345 + 267$?

Your Answer:

5. A town has 8,205 people. During the summer, 2,738 people leave for vacation. How many people are still in the town?

(A) 5,367

(B) 5,467

(C) 5,477

(D) 5,567

6. Estimate $358 + 246$ by rounding each number to the nearest hundred.

Your Answer:

Find more at
ViewMath.com/AZ-Grade3

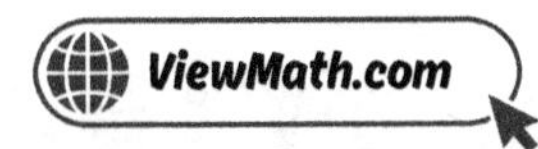

7. *All products of 5 end in which digits?*

   (A) 0 or 2   (B) 0 or 5

   (C) 5 only   (D) 0 only

8. *Which is the largest product you can make with two single-digit numbers?*

   (A) 72   (B) 81

   (C) 90   (D) 100

9. *What is $72 \div 9$?*

   (A) 7   (B) 8

   (C) 9   (D) 63

10. *Find the missing number. $? \times 7 = 63$*

    (A) 7   (B) 8

    (C) 9   (D) 10

11. *Looking at the diagonal of a multiplication table $(1 \times 1, 2 \times 2, 3 \times 3, \ldots)$, what are these numbers called?*

    (A) Even numbers   (B) Odd numbers

    (C) Square numbers   (D) Prime numbers

12. *What is $2 \times 5 + 3 \times 3$?*

    (A) 13   (B) 19

    (C) 39   (D) 45

Find more at
ViewMath.com/AZ-Grade3

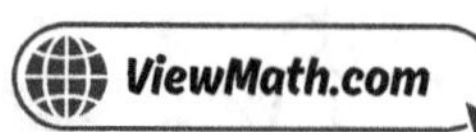

13. *A pie is cut into 6 equal slices. 5 slices are left. What fraction of the pie is left?*

(A) $\frac{1}{6}$

(B) $\frac{5}{6}$

(C) $\frac{6}{5}$

(D) $\frac{6}{6}$

14. *Which fraction is closer to 1 on a number line divided into 6 parts?*

(A) $\frac{2}{6}$

(B) $\frac{3}{6}$

(C) $\frac{4}{6}$

(D) $\frac{5}{6}$

15. *Alex says $\frac{3}{4}$ is 3 copies of $\frac{1}{3}$. Is Alex correct?*

(A) Yes

(B) No, $\frac{3}{4}$ is 3 copies of $\frac{1}{4}$

(C) No, $\frac{3}{4}$ is 4 copies of $\frac{1}{3}$

(D) No, $\frac{3}{4}$ is 3 copies of $\frac{3}{3}$

16. *$\frac{1}{2} = \frac{?}{6}$. What is the missing number?*

(A) 1

(B) 2

(C) 3

(D) 4

17. *Write 4 as a fraction with denominator 3.*

(A) $\frac{4}{3}$

(B) $\frac{3}{4}$

(C) $\frac{7}{3}$

(D) $\frac{12}{3}$

18. *Is $\frac{5}{8}$ more or less than $\frac{1}{2}$? Explain.*

Your Answer:

Find more at
ViewMath.com/AZ-Grade3

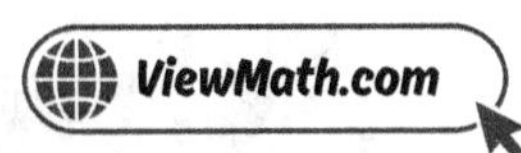

19. Write the time for "quarter past 11" using numbers.

Your Answer:

20. A marker reaches from 0 to the half-inch mark past 6 inches. How long is the marker?

Your Answer:

21. A melon has a mass of 2 kg. A bunch of grapes has a mass of 500 g. How much heavier is the melon than the grapes?

(A) 498 g

(B) 1,000 g

(C) 1,500 g

(D) 2,500 g

22. Container A holds 20 liters. Container B holds 13 liters. How much more does Container A hold?

(A) 7 L

(B) 13 L

(C) 20 L

(D) 33 L

23. Liam buys an apple for $0.85 and a banana for $0.40. He pays with a $5 bill. How much change does he get?

(A) $3.25

(B) $3.75

(C) $4.15

(D) $4.60

24. Subtract: $12.00 − $7.45

Your Answer:

Find more at
ViewMath.com/AZ-Grade3

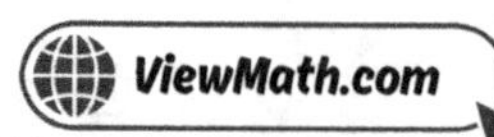

25. Using the same rainy days graph (Jan = 6, Feb = 4, Mar = 3, Apr = 7), how many fewer rainy days were in February than in April?

(A) 1

(B) 2

(C) 3

(D) 4

26. Leo measured 8 worms. His line plot shows 7 X marks total. What went wrong?

(A) He measured one worm twice

(B) He forgot to plot one measurement

(C) He used the wrong numbers on the number line

(D) Nothing — 7 is correct

27. A polygon has 6 sides. What is the name of this polygon?

Your Answer:

28. A square has sides of 9 inches. What is the area?

Your Answer:

29. A triangle has sides of 4 cm, 6 cm, and 8 cm. What is the perimeter?

(A) 14 cm

(B) 18 cm

(C) 20 cm

(D) 24 cm

30. You fold a piece of paper in half, then fold it in half again. How many equal parts do you have?

Your Answer:

 **End of Practice Test 2** 

Great job finishing the test!

 My Score

I got _____________ out of 30 questions right.

*Check your answers in the **Answer Key** at the back of the book.*

Review any questions you missed. That's how we learn!

📊 Check Your Score Online!

Visit **ViewMath Academy** to enter your answers and see which topics you need to review. You can also explore lessons, take quizzes, track your scores, and save your progress!

viewmath.com/score/3.1.AZ.02

Or go to viewmath.com/score and enter code: 3.1.AZ.02

Practice Test 3

 30 Questions

✏ Before You Start ✏

- ✓ **Read each question carefully** before choosing your answer.
- ✓ **Show your work** on scratch paper when you need to.
- ✓ **Skip hard questions** and come back to them later.
- ✓ **Check your answers** when you're done.
- ✓ **Take your time** — there's no rush!

★ You've Got This! ★

Do your best and show what you know!

1. Which number is the same as 9 thousands, 0 hundreds, 2 tens, and 5 ones?

   (A) 9,250

   (B) 9,025

   (C) 9,205

   (D) 9,520

2. When you round 549 to the nearest 10 and to the nearest 100, which gives the larger answer?

   (A) Rounded to the nearest 10

   (B) Rounded to the nearest 100

   (C) They are the same

   (D) Cannot tell without rounding

3. If you add two odd numbers, the result is always:

   (A) Odd

   (B) Even

   (C) Greater than 10

   (D) An odd number less than 20

4. A store sold 256 books on Monday and 478 books on Tuesday. How many books were sold in all?

   (A) 624

   (B) 634

   (C) 724

   (D) 734

5. A school raised $7,500 for charity. They spent $4,825 on supplies. How much money is left?

   Your Answer

6. Which rounding method gives a closer estimate for $463 + 219$?

   (A) Rounding to the nearest 10

   (B) Rounding to the nearest 100

   (C) Both give the same estimate

   (D) You cannot estimate this sum

Find more at
ViewMath.com/AZ-Grade3

7. *What is 2×7?*

   (A) 12
   (B) 14
   (C) 16
   (D) 9

8. *What is 9×6?*

   (A) 45
   (B) 48
   (C) 54
   (D) 56

9. *What is $35 \div 5$?*

   (A) 5
   (B) 6
   (C) 7
   (D) 8

10. *Find the missing number.* $8 \times ? = 48$

   (A) 5
   (B) 6
   (C) 7
   (D) 8

11. *The rule for a pattern is "multiply by 3." If the first number is 2, what are the next two numbers?*

   (A) 5, 8
   (B) 4, 8
   (C) 6, 12
   (D) 6, 18

12. *Solve:* $6 \times 2 + 3 \times 4$

   Your Answer:

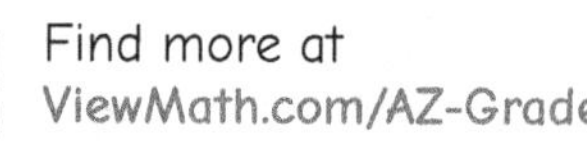
Find more at
ViewMath.com/AZ-Grade3

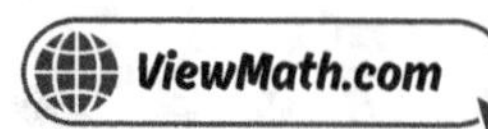

13. A shape is divided into parts, but the parts are NOT equal sizes. Can you write a fraction for the shaded part?

(A) Yes, just count the shaded parts

(B) Yes, fractions work with any parts

(C) No, fractions only work when parts are equal

(D) No, you need to shade all the parts first

14. A number line is divided into 4 equal parts. Which fraction is at the halfway point between 0 and 1?

(A) $\frac{1}{4}$

(B) $\frac{2}{4}$

(C) $\frac{3}{4}$

(D) $\frac{4}{4}$

15. Count by fourths: $\frac{1}{4}, \frac{2}{4}, \underline{\quad}, \frac{4}{4}$. What fraction is missing?

Your Answer:

16. $\frac{4}{8}$ simplified is:

(A) $\frac{1}{4}$

(B) $\frac{2}{4}$

(C) $\frac{1}{2}$

(D) $\frac{2}{8}$

17. Which of these fractions equals 1?

(A) $\frac{3}{4}$

(B) $\frac{6}{6}$

(C) $\frac{4}{6}$

(D) $\frac{1}{1}$ and $\frac{6}{6}$

18. Ruby says $\frac{1}{8} > \frac{1}{3}$ because $8 > 3$. Is Ruby correct?

(A) Yes, bigger numbers are always greater

(B) No, $\frac{1}{3} > \frac{1}{8}$ because thirds are bigger pieces than eighths

(C) They are equal

(D) Yes, eighths are bigger than thirds

Find more at
ViewMath.com/AZ-Grade3

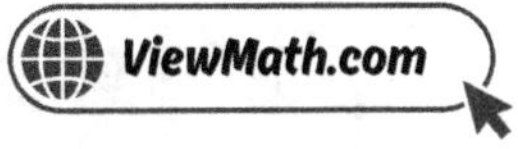

19. The short hand is past the 7 and the long hand is on the 10, then 2 more tick marks past the 10. What time is it?

   (A) 7:50

   (B) 7:52

   (C) 10:37

   (D) 7:10

20. A ribbon is $5\frac{1}{2}$ inches long. A piece of string is $4\frac{3}{4}$ inches long. Which is longer?

   (A) The ribbon

   (B) The string

   (C) They are the same length

   (D) Not enough information

21. Convert 3 kilograms to grams.

   Your Answer:

22. What unit do we use to measure liquid volume?

   (A) Grams

   (B) Kilograms

   (C) Liters

   (D) Inches

23. How many cents are in $1.00?

   (A) 1 cent

   (B) 10 cents

   (C) 50 cents

   (D) 100 cents

24. What is $4.60 + $3.85?

   (A) $7.45

   (B) $8.45

   (C) $8.35

   (D) $7.85

Find more at
ViewMath.com/AZ-Grade3

25. A bar graph has a scale that counts by 10s. The bar for Red reaches the 3rd line. How many items does Red show?

> Your Answer

26. Which type of graph is BEST for showing measurement data like the lengths of worms in inches?

(A) Picture graph   (B) Bar graph

(C) Line plot   (D) Tally chart

27. Which shape is both a rectangle AND a rhombus?

(A) Trapezoid   (B) Pentagon

(C) Square   (D) Triangle

28. A rectangle is 10 feet long and 3 feet wide. What is its area?

(A) 13 sq ft   (B) 26 sq ft

(C) 30 sq ft   (D) 33 sq ft

29. Which shape has the greatest perimeter?

(A) A square with sides of 5 cm   (B) A rectangle that is 8 cm × 3 cm

(C) A rectangle that is 6 cm × 4 cm   (D) A triangle with sides 7 cm, 7 cm, and 7 cm

30. A square is partitioned into 4 equal parts. All 4 parts are shaded. What fraction is shaded?

(A) $\frac{1}{4}$ (B) $\frac{3}{4}$

(C) $\frac{4}{4}$ (D) $\frac{4}{1}$

Find more at
ViewMath.com/AZ-Grade3

 # End of Practice Test 3 

Great job finishing the test!

 **My Score**

I got _____________ out of 30 questions right.

*Check your answers in the **Answer Key** at the back of the book.*

💡 *Review any questions you missed. That's how we learn!*

📊 Check Your Score Online!

Visit **ViewMath Academy** to enter your answers and see which topics you need to review. You can also explore lessons, take quizzes, track your scores, and save your progress!

viewmath.com/score/3.1.AZ.03

Or go to viewmath.com/score and enter code: 3.1.AZ.03

Answer Key & Explanations

 **Check Your Answers!** 

First try each test on your own, then look here to check.

Read the explanations to learn from any mistakes

Practice Test 1 — Answer Key

1	B	2	C	3	B	4	B	5	C	6	800	7	12	8	C	9	72	10	C

| 11 | B | 12 | C | 13 | 6 | 14 | B | 15 | $\frac{5}{6}$ | 16 | C | 17 | B | 18 | Less | 19 | D | 20 | B |

| 21 | C | 22 | 3 L | 23 | D | 24 | B | 25 | D | 26 | C | 27 | Rectangle | 28 | D | 29 | B |

| 30 | B |

💡 Time to Learn! 💡

Go through the explanations below, **especially for the questions you missed.**

Understanding why each answer is correct makes you a stronger math thinker!

👍 **Tip:** Circle any questions you got wrong, then read their explanation carefully.

📖 Practice Test 1 — Detailed Explanations

1. In 7,431: the 7 is in the thousands place and the 4 is in the hundreds place.

Find more at
ViewMath.com/AZ-Grade3

2 462 has tens digit $6 \geq 5$, so it rounds up to 500. 428 and 449 round to 400. 551 rounds to 600.

3 20 (ends in 0), 46 (ends in 6), 88 (ends in 8) are all even. The other groups each contain at least one odd number.

4 The correct answer is $276 + 358 = 634$. Ones: $6 + 8 = 14$, carry 1. Tens: $7 + 5 + 1 = 13$, carry 1. Hundreds: $2 + 3 + 1 = 6$. Marcus got 534, meaning his hundreds digit is 5 instead of 6. He likely forgot to add the carry from the tens column to the hundreds.

5 To check subtraction, add the difference and the number you subtracted: $3{,}627 + 2{,}785$ should equal 6,412.

6 $248 \approx 200$, $175 \approx 200$, and $362 \approx 400$. So $200 + 200 + 400 = 800$.

7 $4 \times 3 = 12$ tires. You can also add: $4 + 4 + 4 = 12$.

8 $48 = 6 \times 8 = 8 \times 6$, so 48 is a multiple of both 6 and 8. Choice A is only a multiple of 6, and choice D is only a multiple of 8.

9 If the number $\div 8 = 9$, then the number $= 9 \times 8 = 72$.

10 Think: $8 \times ? = 72$. Since $8 \times 9 = 72$, the missing number is 9.

11 Odd $\times$ odd always gives an odd product. Example: $3 \times 5 = 15$, $7 \times 9 = 63$.

12 Divide first: $18 \div 3 = 6$. Then add: $9 + 6 = 15$.

13 The denominator tells the total equal parts. The cake is cut into 6 equal parts.

14 $\frac{3}{3} = 1$, so $\frac{6}{3} = 2$.

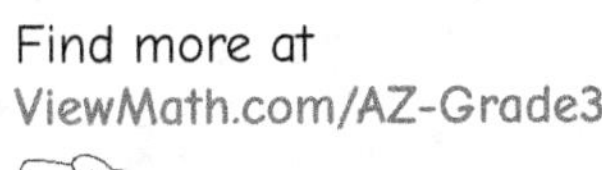

Find more at
ViewMath.com/AZ-Grade3

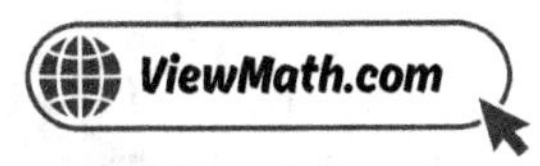

15 $\frac{3}{6} + \frac{1}{6} + \frac{1}{6} = \frac{5}{6}$.

16 The denominator went from 4 to 8 (multiplied by 2). So multiply the numerator by 2 too: $3 \times 2 = 6$.

17 When the numerator is 0, the fraction equals 0. $\frac{0}{4} = 0$.

18 $\frac{1}{2} = \frac{3}{6}$. Since $\frac{2}{6} < \frac{3}{6}$, it is less than $\frac{1}{2}$.

19 When the minute hand points to 12, it means 0 minutes (o'clock). The hour hand on 5 means the hour is 5. The time is 5:00.

20 $6\frac{3}{4} - 6\frac{1}{4} = \frac{3}{4} - \frac{1}{4} = \frac{2}{4} = \frac{1}{2}$ inch.

21 $3 + 5 = 8$ kg.

22 $4 - 1 = 3$ L.

23 A quarter is worth 25 cents.

24 $\$10.00 - \$4.50 = \$5.50$.

25 The scale counts by 5s, so the 4th line is $4 \times 5 = 20$.

26 Crayons shorter than 3 inches: 2 in. (2) + $2\frac{1}{2}$ in. (3) = 5 crayons.

27 It has 4 right angles and 2 pairs of equal sides (but not all 4 sides are equal), so it is a rectangle.

28 A square is a rectangle with all sides equal. Area $= 7 \times 7 = 49$ sq in.

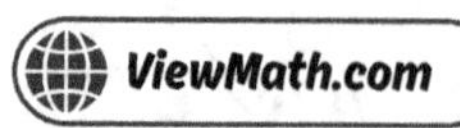

29 Lily multiplied $5 \times 3 = 15$, which gives the area, not the perimeter. The correct perimeter is $5 + 3 + 5 + 3 = 16$ cm.

30 The $\frac{1}{4}$ slices are bigger because the pizza is cut into fewer pieces. More equal parts means smaller pieces: $\frac{1}{4} > \frac{1}{8}$.

Practice Test 2 — Answer Key

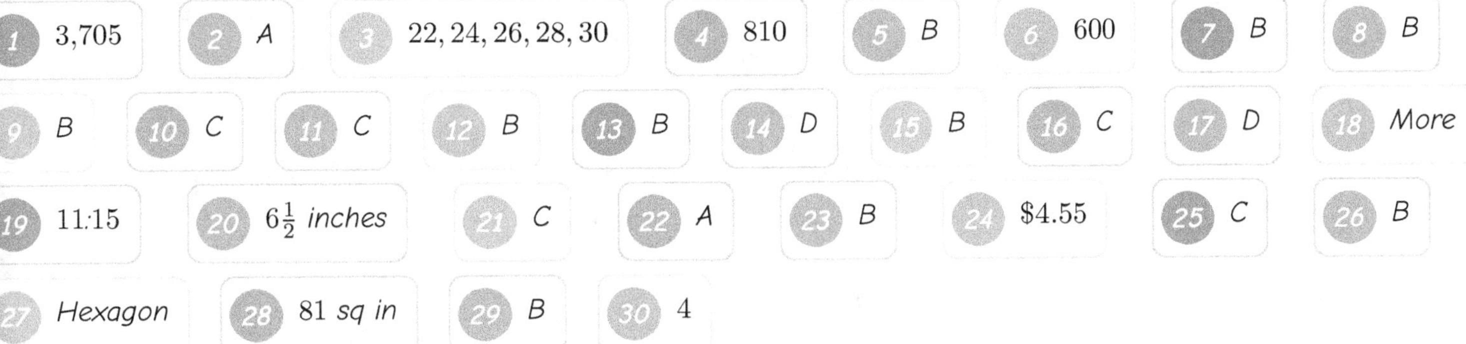

1 3,705	**2** A	**3** 22, 24, 26, 28, 30	**4** 810	**5** B	**6** 600	**7** B	**8** B
9 B	**10** C	**11** C	**12** B	**13** B	**14** D	**15** B	**16** C	**17** D	**18** More
19 11:15	**20** $6\frac{1}{2}$ inches	**21** C	**22** A	**23** B	**24** $4.55	**25** C	**26** B
27 Hexagon	**28** 81 sq in	**29** B	**30** 4

💡 Time to Learn! 💡

Go through the explanations below, **especially for the questions you missed.**

Understanding why each answer is correct makes you a stronger math thinker!

👍 **Tip:** Circle any questions you got wrong, then read their explanation carefully.

📖 Practice Test 2 — Detailed Explanations

1 3 thousands $= 3,000$, 7 hundreds $= 700$, 0 tens $= 0$, 5 ones $= 5$. The number is $3,000 + 700 + 5 = 3,705$.

2 The ones digit is 8. Since $8 \geq 5$, we round up. 438 rounds to 440, not 430.

3. The even numbers between 21 and 30 are $22, 24, 26, 28, 30$. They all end in an even digit.

4. First add $198 + 345 = 543$. Ones: $8 + 5 = 13$, carry 1. Tens: $9 + 4 + 1 = 14$, carry 1. Hundreds: $1 + 3 + 1 = 5$. Then $543 + 267 = 810$. Ones: $3 + 7 = 10$, carry 1. Tens: $4 + 6 + 1 = 11$, carry 1. Hundreds: $5 + 2 + 1 = 8$.

5. $8{,}205 - 2{,}738 = 5{,}467$. Ones: $5 < 8$, borrow through zero: $15 - 8 = 7$. Tens: $9 - 3 = 6$. Hundreds: $1 - 7$, borrow: $11 - 7 = 4$. Thousands: $7 - 2 = 5$.

6. $358 \approx 400$ and $246 \approx 200$. So $400 + 200 = 600$.

7. Products of 5 always end in 0 or 5: $5, 10, 15, 20, 25, 30, \ldots$

8. The largest single-digit number is 9. $9 \times 9 = 81$ is the largest product of two single-digit numbers.

9. Think: $9 \times ? = 72$. Since $9 \times 8 = 72$, the answer is 8.

10. $9 \times 7 = 63$. The missing number is 9.

11. Numbers like $1, 4, 9, 16, 25, 36, \ldots$ where a number is multiplied by itself are called square numbers.

12. Do both multiplications first: $2 \times 5 = 10$ and $3 \times 3 = 9$. Then add: $10 + 9 = 19$.

13. 5 slices out of 6 equal slices $= \frac{5}{6}$.

14. $\frac{5}{6}$ is only $\frac{1}{6}$ away from 1, closer than all the other choices.

15. $\frac{3}{4}$ means 3 copies of $\frac{1}{4}$, not $\frac{1}{3}$. The denominator tells the size of each piece.

Find more at
ViewMath.com/AZ-Grade3

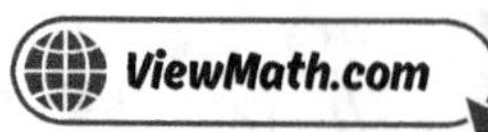

16 The denominator went from 2 to 6 (multiplied by 3). Multiply the numerator too: $1 \times 3 = 3$. So $\frac{1}{2} = \frac{3}{6}$.

17 $4 \times 3 = 12$ thirds. So $4 = \frac{12}{3}$.

18 $\frac{1}{2} = \frac{4}{8}$. Since $\frac{5}{8} > \frac{4}{8}$, it is more than $\frac{1}{2}$.

19 "Quarter past" means 15 minutes after the hour. Quarter past 11 is 11:15.

20 The marker ends at the half-inch mark past 6, which is $6\frac{1}{2}$ inches.

21 2 kg $= 2{,}000$ g. Then $2{,}000 - 500 = 1{,}500$ g.

22 $20 - 13 = 7$ L.

23 Total: $\$0.85 + \$0.40 = \$1.25$. Change: $\$5.00 - \$1.25 = \$3.75$.

24 $\$12.00 - \$7.45 = \$4.55$. Borrow to subtract 45 cents from 0 cents.

25 $7 - 4 = 3$ fewer rainy days in February than April.

26 He measured 8 worms but only has 7 X marks, so he forgot to plot one measurement.

27 A polygon with 6 sides is called a hexagon. "Hex" means 6.

28 Area $= 9 \times 9 = 81$ sq in.

29 Add all side lengths: $4 + 6 + 8 = 18$ cm.

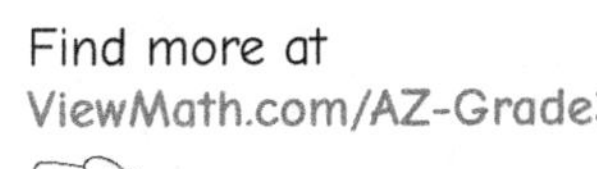
Find more at
ViewMath.com/AZ-Grade3

30 Folding in half gives 2 parts. Folding in half again doubles it to 4 equal parts. Each part is $\frac{1}{4}$ of the whole.

✅ Practice Test 3 — Answer Key

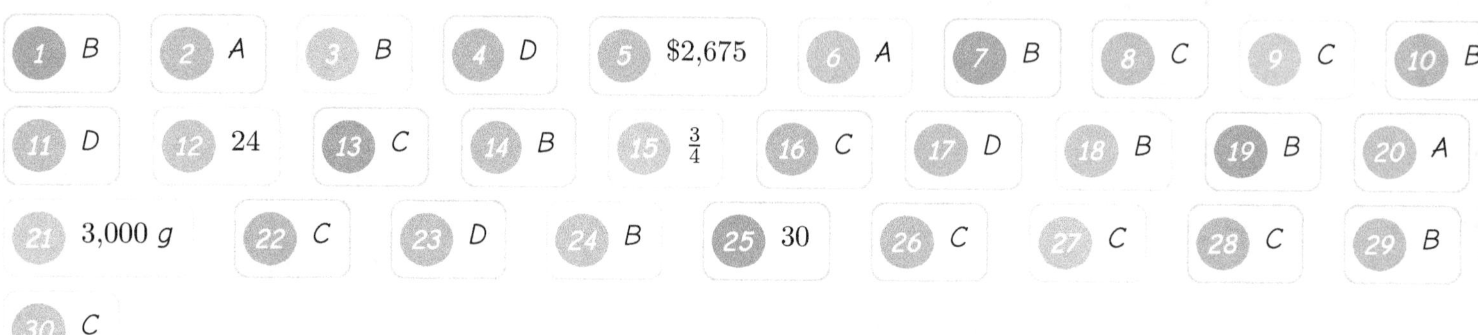

1	B	2	A	3	B	4	D	5	$2,675	6	A	7	B	8	C	9	C	10	B
11	D	12	24	13	C	14	B	15	$\frac{3}{4}$	16	C	17	D	18	B	19	B	20	A
21	3,000 g	22	C	23	D	24	B	25	30	26	C	27	C	28	C	29	B		
30	C																		

💡 Time to Learn! 💡

Go through the explanations below, **especially for the questions you missed**.

Understanding why each answer is correct makes you a stronger math thinker!

👍 **Tip:** Circle any questions you got wrong, then read their explanation carefully.

📖 Practice Test 3 — Detailed Explanations

1 9 thousands $= 9,000$, 0 hundreds $= 0$, 2 tens $= 20$, 5 ones $= 5$. So $9,000 + 20 + 5 = 9,025$.

2 549 rounded to the nearest 10: ones digit is $9 \geq 5$, so 550. Rounded to the nearest 100: tens digit is $4 < 5$, so 500. $550 > 500$.

3 Odd $+$ Odd $=$ Even. For example, $3 + 5 = 8$ (even) and $7 + 9 = 16$ (even).

Find more at
ViewMath.com/AZ-Grade3

4 $256 + 478 = 734$. Ones: $6 + 8 = 14$, carry 1. Tens: $5 + 7 + 1 = 13$, carry 1. Hundreds: $2 + 4 + 1 = 7$.

5 $7,500 - 4,825 = 2,675$. Borrow as needed: ones $10 - 5 = 5$, tens $9 - 2 = 7$, hundreds $4 - 8$ requires borrowing: $14 - 8 = 6$, thousands $6 - 4 = 2$.

6 Nearest 10: $460 + 220 = 680$. Nearest 100: $500 + 200 = 700$. The exact answer is 682, so rounding to the nearest 10 gives a closer estimate.

7 $2 \times 7 = 7 + 7 = 14$. Multiplying by 2 is the same as doubling.

8 $9 \times 6 = 54$. Using the 9s trick: $10 \times 6 = 60$, minus $6 = 54$.

9 Skip count by 5s: $5, 10, 15, 20, 25, 30, 35$. That's 7 jumps. So $35 \div 5 = 7$.

10 $8 \times 6 = 48$. The missing number is 6.

11 $2 \times 3 = 6$, then $6 \times 3 = 18$. The next two numbers are $6, 18$.

12 Do both multiplications first: $6 \times 2 = 12$ and $3 \times 4 = 12$. Then add: $12 + 12 = 24$.

13 Fractions only work when the parts are **equal**. If the pieces are different sizes, you cannot write a fraction.

14 $\frac{2}{4}$ is at the second of 4 marks, which is the middle of the line from 0 to 1.

15 Counting by $\frac{1}{4}$: $\frac{1}{4}, \frac{2}{4}, \frac{3}{4}, \frac{4}{4}$.

16 Divide top and bottom by 4: $\frac{4 \div 4}{8 \div 4} = \frac{1}{2}$.

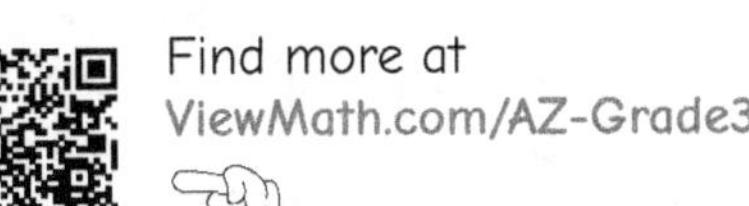
Find more at
ViewMath.com/AZ-Grade3

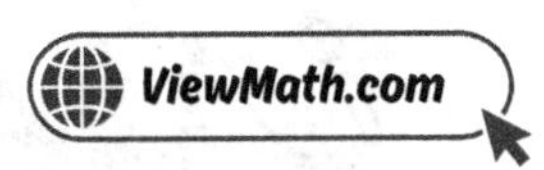

17 Both $\frac{1}{1}$ and $\frac{6}{6}$ equal 1 because numerator = denominator.

18 With the same numerator, the bigger denominator gives **smaller** pieces. $\frac{1}{3} > \frac{1}{8}$.

19 The hour is 7. The long hand on 10 means 50 minutes. Two more tick marks gives $50 + 2 = 52$ minutes. The time is 7:52.

20 $5\frac{1}{2} = 5\frac{2}{4}$, which is greater than $4\frac{3}{4}$. The ribbon is longer.

21 $3 \times 1,000 = 3,000$ g.

22 We measure liquid volume in liters (L).

23 $\$1.00 = 100$ cents.

24 Cents: $60 + 85 = 145$ cents $= 1$ dollar and 45 cents. Dollars: $4 + 3 + 1 = 8$. Answer: $\$8.45$.

25 The scale counts by 10s, so the 3rd line is $3 \times 10 = 30$ items.

26 Line plots are best for measurement data. They show each individual measurement as an X on a number line, making it easy to see how values are spread out.

27 A square has 4 right angles (like a rectangle) AND 4 equal sides (like a rhombus). So a square is both a rectangle and a rhombus.

28 Area $= 10 \times 3 = 30$ sq ft.

29 Square: $4 \times 5 = 20$ cm. Rectangle 8×3: $(2 \times 8) + (2 \times 3) = 22$ cm. Rectangle 6×4: $(2 \times 6) + (2 \times 4) = 20$ cm. Triangle: $7 + 7 + 7 = 21$ cm. The 8×3 rectangle has the greatest perimeter of 22 cm.

 **30** All 4 out of 4 parts are shaded, so $\frac{4}{4}$ is shaded. $\frac{4}{4} = 1$ whole.

Great job checking your work!

Keep practicing and you'll be a math star!

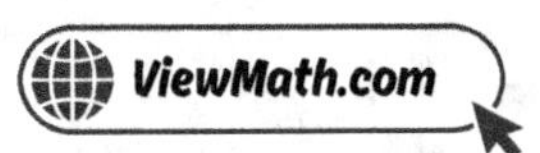